Ask Me Anything About Dinosaurs

Other Avon Camelot Books by
Louis Phillips

ASK ME ANYTHING ABOUT BASEBALL
ASK ME ANYTHING ABOUT MONSTERS
ASK ME ANYTHING ABOUT THE PRESIDENTS

LOUIS PHILLIPS, author of more than thirty books for children and adults, has been collecting facts on just about everything his entire life, and this love for odd and unusual information has remained with him to this day.

Mr. Phillips lives in New York City with his wife and twin sons.

Avon Books are available at special quantity discounts for bulk purchases for sales promotions, premiums, fund raising or educational use. Special books, or book excerpts, can also be created to fit specific needs.

For details write or telephone the office of the Director of Special Markets, Avon Books, Dept. FP, 1350 Avenue of the Americas, New York, New York 10019, 1-800-238-0658.

Ask Me Anything About Dinosaurs

LOUIS PHILLIPS

Illustrations by Kevin Wasden

AN AVON  CAMELOT BOOK

AVON BOOKS
A division of
The Hearst Corporation
1350 Avenue of the Americas
New York, New York 10019

Copyright © 1997 by Louis Phillips
Interior illustrations copyright © 1997 by Avon Books
Interior illustrations by Kevin Wasden
Published by arrangement with the author
Visit our website at **http://AvonBooks.com**
Library of Congress Catalog Card Number: 96-48486
ISBN: 0-380-78552-8
RL: 5.8

All rights reserved, which includes the right to reproduce this book or portions thereof in any form whatsoever except as provided by the U.S. Copyright Law. For information address Fifi Oscard Agency, Inc., 24 West 40th Street, 17th floor, New York, New York 10018.

Library of Congress Cataloging in Publication Data:

Phillips, Louis.
  Ask me anything about dinosaurs / Louis Phillips.
    p.   cm.
Summary: Questions and answers present information about dinosaurs from different periods.
1. Dinosaurs—Miscellanea—Juvenile literature. [1. Dinosaurs—Miscellanea. 2. Questions and answers.] I. Title.
QE862.D5P494 1997   96-48486
567.9—dc21   CIP

First Avon Camelot Printing: June 1997

CAMELOT TRADEMARK REG. U.S. PAT. OFF. AND IN OTHER COUNTRIES, MARCA REGISTRADA, HECHO EN U.S.A.

Printed in the U.S.A.

OPM 10 9 8 7 6 5 4 3 2 1

> If you purchased this book without a cover, you should be aware that this book is stolen property. It was reported as "unsold and destroyed" to the publisher, and neither the author nor the publisher has received any payment for this "stripped book."

For Jonathan, Janice, & Ben Sternberg—
a new home deserves a new book

ACKNOWLEDGMENTS

A work of this nature could not have been written without the research of thousands of paleontologists and without access to scientific writings in journals and in popular magazines and newspapers. The author gratefully acknowledges the library and staff of the American Museum of Natural History in New York City, and the libary and staff of the New York Public Library. Thank you also to Joan Kramer, who helped track down numerous names and addresses. The author also wishes to acknowledge the encouragement and support provided by his patient editor at Avon Books, Gwen Montgomery, and the superb staff of copyeditors and artists. Any errors that may have crept unknowingly, unwittingly, and foolishly into this text are the fault of the author alone.

What were the largest animals ever to walk upon the earth?

Dinosaurs.

What is the oldest dinosaur that we have record of?

The oldest dinosaur that we have thus far recovered from the past would most likely be *Chindesaurus,* a long-necked, long-tailed reptile that lived about 225 million years ago. *Chindesaurus* stood about three feet high and weighed approximately two hundred pounds.

Chindesaurus is named for that part of the Arizona desert called Chinde Point. *Chinde* is a Native American word that means "ghost." Hence, *Chindesaurus* means "ghost lizard." Coming across one of them would be enough to frighten anyone.

What was the first *large* dinosaur to walk the earth?

The first large dinosaur to walk the earth was most likely *Plateosaurus* ("flat lizard"). It grew to be about twenty-five feet long and could stand on its hind legs (using its tail for balance) to eat leaves from tall trees. Its hand had four fingers and a clawlike thumb which might have been used to rake ground plants or to tear leaves from trees. In addition, the clawlike thumb was undoubtedly used as a weapon or a means of defense. The remains of a *Plateosaurus* were first described in 1837 by the scientist Hermann von Meyers. Hermann von Meyers, an early German paleontologist, also described *Archeopteryx* (1861) and *Stenopelix* (1857).

Why is the term *dinosaur* not very accurate?

Dinosaur means "terrible lizard," but dinosaurs are reptiles, and not lizards at all.

What does the name *sauropod* mean?

Well, we know that *sauro* means "lizard" and *pod* refers to "foot." Thus, *sauropod* means "lizard foot." The family of dinosaurs known as sauropods received that name because the feet of a sauropod resembled the feet of modern-day lizards.

What kinds of large dinosaurs belong to the family of sauropods?

Among the members of the sauropod family are *Apatosaurus, Brachiosaurus, Camarasaurus, Diplodocus,* and two very large dinosaurs that have been nicknamed *Supersaurus* and *Ultrasaurus.*

Were dinosaurs the first animals to breathe air and walk on dry land?

No, not by a long shot. The first animals to breathe air and walk on land were scorpions. The first vertebrates to walk on dry land were the amphibians, which emerged from the oceans over 408 million years ago.

How do dinosaurs get their names?

The right to name a fossil usually falls to the scientist who discovers it. However, in order to promote clear communication among scientists around the world, the naming of dinosaurs always follows certain guidelines. First, the scientific name will always be in latinized Greek or Latin, and the name will be in two parts: the first part relating to the genus and the second part to the species.

The common, or trivial name, however, may be built from a word that expresses some real or imagined aspect of the animal to which it applies. The name *Anatosaurus,* for example, means "goose lizard," although it is commonly referred to as a duck-billed dinosaur. *Ornitholestes,* one member of the family of coelurosaurs, received its name, meaning "bird catcher," because it was once believed that this dinosaur leaped into the air to catch low-flying birds. This belief, however, is no longer held in high regard. It sometimes happens that, because of a misunderstanding, a false belief, or ignorance of Latin, creatures receive names that do not fit their true attributes.

Protoceratops means "first horned face," but it has no horns. In naming this dinosaur, scientists realized that *Protoceratops* was related to a family of dinosaurs that *did* have horns: the ceratopsian, or "horned face" fam-

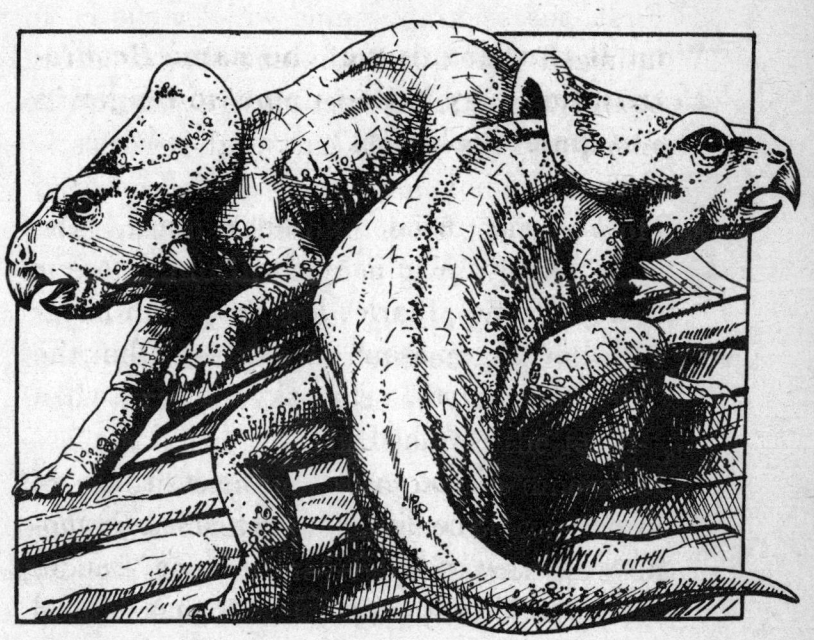

ily. *Protoceratops* was probably the first of this line, and so the name "first horned face" is more appropriate than it might seem.

Sometimes the name is a compound word, perhaps combining a classic Latin root with a proper name. *Lambeosaurus* consists of *Lambe* (after Lawrence Lambe, a Canadian paleontologist) and *sauros* ("lizard"). A name that combines a classic Latin root with a non-Latin root is called a *barbarous* name.

What is the meaning of the name *Brontosaurus*, and why is that name no longer in use by paleontologists?

Brontosaurus means "thunder lizard." The name was given to it by the great dinosaur hunter Othniel Charles Marsh. Marsh imagined that the dinosaur was so large that the earth "thundered," or shook violently, when *Brontosaurus* walked.

The name *Brontosaurus,* however, is now an invalid term. The dinosaur once called *Brontosaurus* is now called *Apatosaurus,* or "fraudulent lizard." What happened was that Othniel

Marsh had discovered two sets of fossils and, thinking they were from different dinosaurs, had given them two different names. It eventually came to light that the two dinosaurs were two different species of the same genus. Since *Apatosaurus* had been named first, it became the correct scientific genus name.

There once was a dinosaur that was named *Elosaurus*. This was also a mistaken name. The *Elosaurus* specimen turned out to be a young *Apatosaurus*.

Why did paleontologists change the name of the animal known as *Stereocephalus* ("twin head") to *Euoplocephalus*?

Euoplocephalus was a very large ankylosaur weighing about four thousand pounds. This dinosaur had pointed spikes all the way down its back and on its tail. Lawrence Lambe originally named it *Stereocephalus,* but the name had to be changed when Dr. Lambe discovered that there was an insect already known by this name. The name Euoplocephalus means "well-armed head."

What is a fossil?

A fossil is the remains of some once-living organism (plant or animal) from a past geological age. The fossil—a skeleton, leaf imprint, or footprint—usually dates very far back in geological time and has been preserved in the crust of the earth. The word fossil itself comes from the Latin *fodere,* which means "to dig." This is appropriate, since fossils are often dug up.

Should you ever have the privilege of journeying to the fourth floor of the American Museum of Natural History in New York City, you will be informed that:

More than 2,500 years ago the Greeks comprehended in a general way that fossils are the remains or traces of organisms buried by natural means in the earth. With the advent of the Dark Ages, this ancient understanding of fossils was largely forgotten. For many centuries thereafter, fossils were regarded as special creations originating in the rocks or as objects created in the heavens and

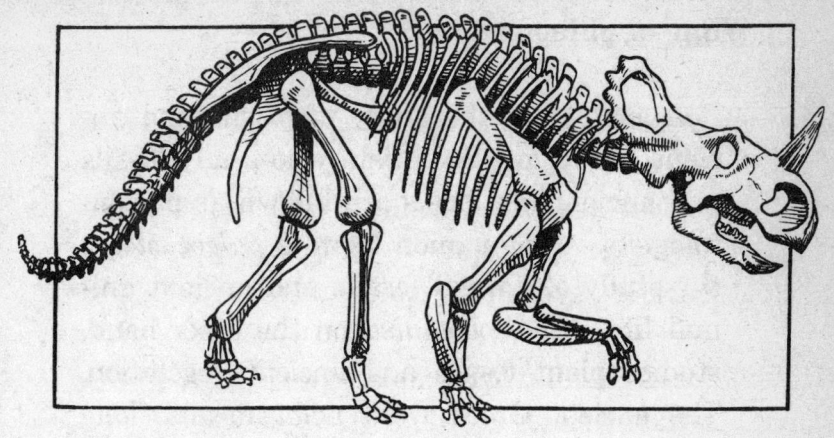

fallen to earth. They were also commonly thought to be the remains of organisms buried during the biblical deluge. During this time Leonardo da Vinci and a few others did interpret fossils correctly, but it was not until the 18th Century that three important ideas involving fossils became firmly established: (A) That there is a succession through time of different fossils in the rocks. (B) That this succession is the record of the history of life, and (C) That fossils can provide the basis for dividing geologic time.

What is paleontology?

Paleontology is the study of fossils and ancient life forms. Scientists who study fossils and ancient life forms are known as paleontologists. A companion term is *paleozoology,* the study of animal fossils and ancient animal life. A *paleobotanist,* on the other hand, studies plant fossils and ancient vegetation. The ancient Greek word *palai* means "long ago." Obviously, the study of dinosaurs takes us into long ago times.

Is it common for paleontologists to find complete fossil skeletons?

No. It is very rare for paleontologists to find entire skeletons. Scientists often have to reconstruct entire skeletons using only a few major bones as clues. It can take many years for an entire skeleton to be assembled for museum exhibition.

Who was the first person credited with discovering a dinosaur fossil?

The Reverend Dr. Robert Plot is credited with being the first person to record the discovery of a dinosaur fossil. In 1676, the Reverend Plot described the lower section of a very large *femur,* or thigh bone. At the time he did not know that it belonged to a dinosaur; he thought that the bone was the thigh bone of a giant human being. Scientists today believe (although the bone discovered by Dr. Plot has long been lost) that the bone was that of *Megalosaurus.*

What are the names of some of the meat-eating dinosaurs known as theropods?

The name "theropod" means "beast feet." Among the members of that beastly-footed family are:

1. tyrannosaurids
2. spinosaurids
3. segnosaurids

4. megalosaurids
5. dromaeosaurids
6. deinonychosaurs
7. ornithomimids
8. procompsognathids
9. coelurosaurs

Did all dinosaurs have tiny brains?

No, not at all. *Dromaesaurus,* for example, had a fairly large brain, and *Stenonychosaurus* had a brain that was at least the size of an ostrich's brain.

What did dinosaurs eat?

Most dinosaurs were eaters of plants and vegetation, but there were a few dinosaurs that ate meat. Theropods, for example, were meat eaters.

Who was Georges, Baron Cuvier, and why is he so important to the study of dinosaurs?

Georges, Baron Cuvier (1769-1832) is sometimes referred to as the "Father of Paleontology" because this French naturalist was the first scientist to classify fossils into distinct families, or orders.

In 1812, he published *Researches sur les ossements fossiles de quadrupedès* (a study of fossils) which is considered to be the earliest systematic account of paleontology.

What great paleontologist was named in honor of the great circus showman P.T. Barnum?

Barnum Brown (1873-1963) received his first name because P.T. Barnum was so popular at the time of Brown's birth. Barnum Brown is known today as the greatest dinosaur hunter of the twentieth century. In fact, he was dubbed the "Father of Dinosaurs" because he led so many expeditions to recover fossils. He discovered dinosaurs in Texas, Oklahoma, Montana, Colorado, Arizona, and many other places in the United States and abroad. In 1934, he was awarded an honorary degree of Doctor of Science by Lehigh University in recognition of his scientific work, including his successful explorations for the fossilized remains of prehistoric mammals and reptiles on the American continent and in Asia, Europe, and Africa. His wife, Lillian Brown, published three books about their travels and adventures together in the field: *I Married a Dinosaur, Cleopatra Slept Here,* and *Bring 'em Back Petrified.* Brown was known for being so meticulous that he even wore a topcoat in the field and dressed quite formally while on a dig. If you should ever go to the American Museum of Natural History in New York City, you will be able to view many of the fossil skeletons and treasures that were brought to the museum by Barnum Brown.

What discovery pertaining to dinosaurs did the great Victorian scientist T.H. Huxley make while carving a turkey on Christmas day?

The following story may or may not be true, but it has been frequently repeated and is certainly memorable: It seems that one Christmas day, not too long after the discovery of *Archaeopteryx* in 1869, Huxley was having Christmas dinner with his family. As he started to carve a drumstick from the turkey, he noticed a startling similarity between the bone structure of the bird and the fossilized bones of a *Megalosaurus* that were back in his office. Huxley therefore came to believe, along with fellow scientist Richard Owen, that the origin of birds is directly related to the evolution of dinosaurs. As Huxley himself noted, ". . . surely there is nothing very wild or illegitimate in the hypothesis that the phylum of the class Aves has its roots in the dinosaurian reptiles."

Today it is widely accepted fact that birds are a form of theropod dinosaur.

Why are Othniel Charles Marsh and Edward Drinker Cope so important to the history of paleontology?

Othniel Charles Marsh, director of the George Peabody Museum's department of paleontology at Yale University, and Edward Drinker Cope, the son of rich Pennsylvania Quaker parents, were the two most prominent paleontologists in nineteenth-century America. Both men discovered and classified thousands of fossilized dinosaur bones. Between them they identified more than eighteen hundred new species or genera of prehistoric life. Unfortunately, the two scientists hated each other, and the rivalry between them was intense. As Tim DeForest tells us in his article "The Search For Monsters" in the February 1995 issue of *Wild West*:

> *Cope and Marsh would both lead expeditions into the unsettled badlands of the West, braving harsh weather, outlaws and hostile Indians as they scoured the land for fossils. Both would spend the bulk of their personal fortunes funding*

massive fossil digs. Both were also monumental egoists, each convinced—not without justification—of his own brilliance. It was this egotism, combined with their passion for fossils, that would inspire mutual hatred and drive them into a rivalry that would span three decades.

Who was James Parkinson, and what contribution did he make to the study of dinosaurs?

James Parkinson was a noted nineteenth-century English physician (the physical disorder known as Parkinson's Disease is indeed named after him). He was also a geologist and, in 1822, published a slender volume titled *Outlines of Oryctology, An Introduction to the study of Fossil Organic Remains especially of those Found in the British Strata*. It was in this book that Dr. Parkinson mentioned a fossil that he had named *Megalosaurus*, which means "great lizard." *Megalosaurus* was the first dinosaur to be given a scientific name. The first detailed description of it, however, did not appear until two years later, and was recorded by a man named William Buckland. It is Buckland who is frequently credited with naming that dinosaur.

How do paleontologists determine the kinds of foods that were eaten by extinct dinosaurs?

The two strongest clues are the contents of fossilized stomach cavities and the shape of dinosaurs' teeth. The teeth of meat-eating dinosaurs, for example, were either very sharp and serrated, or they were daggerlike. Vegetable-eating dinosaurs, however, had teeth that were flattened at the top, or peg-shaped.

How is it possible for scientists to deduce what dinosaurs actually ate?

There are a number of clues that paleontologists use:

1. Type, number, and shape of a dinosaur's teeth
2. Bones of other creatures discovered within dinosaurs' skeletons
3. Teeth marks left on bones of other creatures, indicating that the bones were chewed upon by dinosaurs
4. Contents of fossilized dinosaur excrement or droppings

5. Types of foods available to dinosaurs of particular geographic locations and climates during a specific geological time period

Which kinds of dinosaurs had larger stomachs: the meat-eating dinosaurs or the plant-eating ones?

The stomachs of plant eaters are, by necessity, usually larger than the stomachs of meat eaters. Why? Because it takes a large amount of vegetable and plant materials to produce—during and after the process of digestion—enough nutrients to sustain an animal. Hence, the plant-eating dinosaurs had the larger stomachs.

What ostrichlike dinosaur had a brain that was larger than the brain of any bird living today?

That would be *Struthiomimus* a name that means "like an ostrich." *Struthiomimus* had no teeth, but it had a birdlike beak. Also, *Struthiomimus* could run fairly fast.

If it were possible to hold a foot race between a modern-day ostrich and *Tyrannosaurus rex*, which creature would win?

Most likely the ostrich would outrun *T. rex* (and probably would have to, in order to stay alive!). *Tyrannosaurus rex* could average a running speed of eighteen miles per hour, whereas the modern-day ostrich can run nearly twenty-six miles per hour.

What major paleontological discovery of 1996 was made in China?

In 1996, a fossil was discovered in China that showed traces of feathery down along its spine and sides. The discovery of the feathery fossil was a sign that such a dinosaur (which appeared to be a type of compsognathid carnivore) may have been an early ancestor of birds.

Many of the experts who studied photographs of the fossil stated that this discovery gave further support to the theory that birds descended from dinosaurs.

Did dinosaurs possess a good sense of smell?

Because dinosaurs were most active during daylight, they relied heavily upon their eyesight to locate food. But some dinosaurs, such as *Brachiosaurus,* had skulls that contained large nasal openings. This led some paleontologists to conclude that dinosaurs with large snouts and large nasal openings most likely possessed a good sense of smell.

What dinosaur had the smallest brain in relation to its overall body size?

Stegosaurus had a body weight of nearly two tons, but its brain weighed only two and one-half ounces. If *Stegosaurus* weighed 1.9 tons, for example, its brain would represent only

0.0004 of one percent of its body weight. A human brain, on average, accounts for 1.88 percent of a human's overall body weight. The brain of an average-sized male human being weighs about fifty ounces, the brain of the female weighing a few ounces less.

Can sense be made of the name *Macroelongatoolithus xixiaensis*?

If you wish to impress your friends (and perhaps some of your teachers), it might be fun for you to learn to spell the name *Macroelongatoolithus xixiaensis*. The thirty-letter name looks extremely difficult, but if you break it down into parts, the name can be easily deciphered. The latin prefix *macro* means "large," *elongat* means "elongated", *oo* means "egg", and *lithus* means "stone." *Xixiaensis* refers to the fact that the fossil was discovered in the Xixia Basin of China.

Hence, the scientific name *Macroelongatoolithus xixiaensis* means "a very large elongated fossilized egg found in the Xixia Basin of China."

Who was Dr. Gideon Mantell?

If you should ever travel to Lewes, England (just south of London), you might find yourself standing in front of the house that once belonged to Dr. Gideon Mantell (1790–1852). Fastened to the door of this house is a brass plate that reads:

HE DISCOVERED THE IGUANODON

Iguanadon was the first extinct dinosaur to be discovered in England. Dr. Mantell published a paper about his discovery in 1825.

How did *Iguanodon* get its name?

In 1822, the wife of Dr. Gideon Mantell discovered some strange-looking teeth embedded in some rocks in Sussex, England. Some of the scientists to whom she showed the teeth at first thought that the teeth had belonged to a rhinoceros.

After his wife's discovery, Dr. Mantell returned to the site and found more fossils to go

with the teeth. He then concluded that the teeth and bones belonged to some kind of prehistoric lizard or dinosaur. He named the dinosaur *Iguanodon,* because the teeth resembled the teeth found in the mouths of iguanas!

The next time you look an iguana in the mouth, think of the lowly *Iguanodon.*

In 1993, the Dinosaur Society renamed the oldest known ankylosaur. It is now known as *Jurassosaurus nedegoapeferkimorum*. What is unusual about that dinosaur's second name?

Part of the second name—*nedegoapeferkimorum*—is an acronym consisting of the initials of the last names of the actors and actresses in the movie *Jurassic Park*.

How did the dinosaur known as *Deinonychus* receive a name that means "terrible claw?"

Deinonychus was a fierce meat-eating dinosaur that was about eight to ten feet long and had large muscular legs and clawed toes. It had a large, curved, switchblade-like claw on the second toe of each foot; hence, its name.

Deinonychus lived during the early Cretaceous period in what is now western United States.

What mistake did paleontologists make in naming *Mussaurus*?

When paleontologists discovered young *Mussarus* fossils, they noted how small the creature was and, so designated it the "mouse lizard." The mouse lizard was among the oldest dinosaurs on earth and lived in southern Argentina some 210 million years ago, during the Triassic period.

Scientists later learned that *Mussaurus* was not as tiny as previously believed. Adults of the species actually grew to more than ten feet in length. Thus, it was much larger than a mouse.

There is a very famous children's book about a horse called Black Beauty. In the world of paleontology, what dinosaur skeleton is affectionately referred to as "Black Beauty?"

"Black Beauty" refers to a skeleton of *Tyrannosaurus rex* discovered in Alberta, Canada. While the bones were becoming fossilized, magnesium compounds from the groundwater seeped into them and turned the bones black. Hence, the nickname.

Has any Dinosaur been named in honor of a book?

At least one has. *Huayangosaurus,* discovered in the early 1960s in the Szechuan province of central China was named in honor of an old Chinese text, *Hua Yang Guo Zhi,* a book from the Jin Dynasty (A.D. 265–317). Huayang was an early name for Szechuan.

Huayangosaurus, a small *Stegosaurus* covered with spikes, lived about 170 million years ago.

How did *Heterodontosaurus* get a name that means "different-toothed lizard?"

Heterodontosaurus received its name from the fact that both the females and males of that species possessed two different kinds of teeth. The back teeth were used for chewing,

DINOSAURS IN FICTION

Ahlberg, Allan. *Dinosaur Dreams*. Greenwillow Books, New York, 1991.

Birchman, David Francis. *Brother Billy Bronto's Bygone Blues Band*. Lothrop, Lee & Shepard, New York, 1992.

Blackwood, Mary. *Derek, the Knitting Dinosaur*. Carolrhoda Book, Minneapolis, 1995.

Butterworth, Olive. *The Enormous Egg*. Little Brown & Company, Boston, 1956.

Haynes, Max. *Dinosaur Island*. Lothrop, Lee & Shepard, New York, 1991.

Lasky, Kathryn. *The Bone Wars*. William Morrow, Co., New York, 1988.

Pittman, Helena Clare. *A Dinosaur for Gerald*. Lerner Books, Minneapolis, 1990.

but the teeth in the front of their mouths were used for biting.

What dinosaur was given the nickname "Claws?"

"Claws" is a nickname that some scientists apply to the carnivorous dinosaur called *Baryonyx*. Its skeleton was discovered in England, and it was named in 1986. The hand exhibited a large, curved, hard claw; hence, its nickname. *Baronyx* was about thirty feet long and weighed about two tons. Some paleontologists believe that the *Baronyx* used its claw like a small spear or harpoon to capture fish.

What dinosaur received its name from the group of stars known as the Southern Cross, which forms a cross in the sky over the southern hemisphere?

That would be *Staurikosaurus,* whose name literally means "cross lizard." Some people might think the name refers to its temper or its bad moods, but in fact the dinosaur was named for the Southern Cross over our heads.

What can paleontologists learn from studying dinosaur footprints in trackways?

The trackways preserve the dinosaurs' footprints from which scientists can learn all kinds of information. For example, the length of the stride allows scientists to determine just how fast the dinosaurs were moving. The patterns of the tracks reveal whether certain species of dinosaurs traveled in herds or not, and whether they walked one behind the other or side by side. Did dinosaurs that were injured or walked with a limp keep up with the others? This is just one of the questions a good trackway might answer.

Scientists have discovered air canals in the skulls of *Tyrannosaurus rex*. What were the functions of these air holes?

It has been conjectured that these air canals in *T. rex's* skull were present in order to cool its head. The air canals also reduced the weight of the large head. A third possibility is that the canals helped *T. rex* hear better.

There is a comic book hero named Superman, but what is *Supersaurus*?

In 1922, a scientist named Jim Jensen found a fragmentary skeleton in Colorado. By making estimates based upon the sizes of the neck vertebrae and shoulder bones, Dr. Jensen suggested that this new dinosaur might have stood fifty-four feet tall and had a body length of nearly ninety-eight feet. Because of the dinosaur's great size, he nicknamed his discovery "Supersaurus."

What was the largest single fossil bone ever found?

The largest single fossil bone ever found belonged to a dinosaur that has been nicknamed "Ultrasaurus" (it has yet to receive an official scientific name). The shoulder bone is nine feet long. It was discovered in the state of Colorado in the 1970s.

Some dinosaurs, such as *Apatosaurus* and *Diplodocus*, grew to be very large (some were eighty to ninety feet long and weighed about thirty tons). Although there were many advantages to being so large, what were some of the disadvantages?

Because sauropod necks were so long—sometimes over thirty-five feet—it was very difficult for their blood to reach their heads. Also, the bigger the dinosaurs got, the more food they needed; thus, they became more vulnerable to extinction.

What is an omnivore?

An *omnivore* is an animal that eats both meat and plants.

What kinds of plants did plant-eating dinosaurs eat?

The plant-eating dinosaurs (herbivores) made their vegetarian meals from cycad fronds, horsetails, pine needles, and conifers.
During the first period of the dinosaur's reign, the first flowering plants appeared. Magnolias, for example, appeared about one hundred million years ago.
Some dinosaur lovers might be pleased to imagine their favorite dinosaur eating a meal of magnolias or dogwood.

Did any dinosaurs eat grass?

No. Grass did not appear upon the earth until long after dinosaurs had become extinct.

What was the smallest of all the meat-eating dinosaurs?

That honor—if we may call it that—goes to *Compsognathus* (its name means "pretty jaw"). *Compsognathus* was not much bigger than a modern-day chicken and weighed only a few pounds.

What insects evolved during the Jurassic period, when dinosaurs were at their peak?

During the Jurassic period (which occurred 144 million to 213 million years ago) a number of insect species came into being. Ants, wasps, flies, and bees all evolved during the Jurassic period.

What dinosaur had the longest neck?

That honor would belong to *Mamenchisaurus,* or "Mamenchi lizard," whose body measured over seventy feet. Over half of that length was its neck. Imagine a neck measuring thirty-five to thirty-six feet! *Mamenchisaurus* could make the modern giraffe blush with modesty.

Why did very large dinosaurs, such as the *Hypacrosaurus*, lay small eggs? Shouldn't dinosaur eggs have been huge?

Big eggs are not always very efficient in promoting the hatching of healthy animals. The problem with big eggs is that they have thick shells. If you were an embryo inside an egg you would need more strength to break through a thick shell than you would need to break a thin shell. Also, a very thick shell would make it much more difficult for the embryo inside the egg to get oxygen.

How did the blood pressure of *Apatosaurus* compare with that of a human being?

Well, we know that a modern-day giraffe, when compared to a human being, has very high blood pressure. The pressure is used to pump blood all the way up its long neck to its head. Now, since *Apatosaurus* had a neck even longer than a giraffe's, it is possible to argue, by means of analogy, that *Apatosaurus* must also have had high blood pressure.

What was the fate of the "paleozoic museum" that was to be built in New York City in the late nineteenth century?

In the latter part of the nineteenth century, plans were drawn up to construct a building of iron and glass, modeled (on a much smaller scale) in the style of London's famous Crystal Palace. This new building was to be designated a paleozoic museum, to be located near Eighth Avenue and Sixty-third Street in New York City.

A zoologic artist named Waterhouse Hawkins worked long and hard, along with numerous assistants, to create models of the dinosaurs to be displayed. Hawkins planned to show such creatures as hadrosaurs, plesiosaurs, mastodons, and two armadillo-like glyptodonts (although mastodons and glyptodonts lived long after dinosaurs had become extinct).

Unfortunately, the paleozoic museum was never built, but it is fun to imagine what it might have been like and how its large dinosaur models would compare to today's more accurate reconstructions.

What does an ichnologist study?

Ichnology is the study of footprints preserved in rocks; hence, an ichnologist studies those preserved footprints.

Since so many dinosaurs roamed the earth for such long periods of time, why are there so few dinosaur footprints for ichnologists to study?

The reason so many dinosaur footprints have disappeared is that time and erosion have erased most of them. It is estimated that ninety-nine percent of all the dinosaur footprints have been erased.

What does a paleoscatologist do?

A paleoscatologist is a scientist who studies, analyzes, and categorizes the fossilized excrement of dinosaurs. Fossilized excrement is known as a coprolite. There are not many paleoscatologists in the world. In fact, Karen

Wright, author of "What the Dinosaurs Left Us" (*Discover,* June 1996) states: "It is safe to say that [Karen] Chin is the world's leading paleoscatologist. Also the world's only paleoscatologist."

Ms. Wright goes on to say: "In coprolites, she [Karen Chin] hopes to find evidence of feeding habits and behavior available from no other fossil source. Most important, she expects to discover the diets of ancient creatures so that paleontologists may one day reconstruct ecological webs from the very bowels of prehistory."

What do the words *ectothermic* and *endothermic* mean?

Well, the words are not as complicated as they look. Ectothermic dinosaurs were cold-blooded (like snakes), whereas endothermic dinosaurs were warm-blooded (like birds).

Are you ectothermic (cold-blooded) or endothermic (warm-blooded)?

What is the distinction between a warm-blooded creature and a cold-blooded one?

A warm-blooded creature maintains a relatively constant, warm body temperature independent of environmental temperature. On a cold day or a hot day, the normal body temperature of a human being, for example, is 98.6 degrees Fahrenheit.

A cold-blooded animal is one whose body temperature varies with the external environment. Snakes, for example, are cold-blooded.

Were extinct dinosaurs warm-blooded?

So far there is no clear-cut evidence that will allow scientists to determine whether extinct dinosaurs were warm-blooded or cold-blooded. About a decade ago, many paleontologists favored the theory that dinosaurs were warm-blooded, but their arguments were based on indirect evidence. More recently, scientists have begun studying the nasal cavities of living animals such as crocodiles (cold-blooded) and raccoons (warm-blooded), and comparing them to the nasal cavities of dinosaurs. Warm-blooded animal noses contain sheets of bone or cartilage called turbinates. According to a recent theory by John Ruben, a physiologist at Oregon State University, and Willem Hellenius of the College of Charleston in South Carolina, the answer to this question—were dinosaurs cold-blooded or warm-blooded—lies chiefly in the noses. Because they have less need for oxygen, cold-blooded animals, such as crocodiles, have nasal cavities that are small for their bodies. Dinosaurs, it appears, had small nasal cavities that lack turbinates. This analogy suggests dinosaurs were cold-blooded.

A discussion for the general reader of the nasal-passage theory can be found in Virginia Morrell's article, "A Cold, Hard Look at Dinosaurs," in the December 1996 issue of *Discover* magazine.

Which kind of animal, a cold-blooded one or a warm-blooded one, requires more food?

Warm-blooded birds and mammals require ten times more food than cold-blooded ones.

How can scientists deduce just how smart some dinosaurs were?

Scientists do this by studying the braincases left behind. The gray matter of the brain itself leaves impressions on the bone. From these bone markings, paleontologists infer which sensory regions were developed or not; thus, scientists can make educated guesses about how smart certain dinosaurs were.

What were some uses for the long tails of dinosaurs?

The long tails that many dinosaurs had were used in many different ways. The tails served as anchor posts for the leg muscles. They provided balance. Dinosaurs could rear up and, by balancing on their tails, reach higher up into the trees for food. The tails also provided balance while dinosaurs were running.

The tails also frequently provided an additional means of defense. Sauropods, for example, could have used their tails like whips. The dinosaurs that used their tails in defense were the four-footed herbivorous ones.

What dinosaur—discovered recently in the Sahara Desert in Africa—was, perhaps, the largest carnivore that ever walked upon the earth?

Carcharodontosaurus, or "shark-toothed reptile," was probably the largest carnivore. (It was a bit longer than *Tyrannosaurus rex.*) Its skull was discovered by Paul Serrano, an associate professor of paleontology and evolution at the University of Chicago. The skull measured five feet four inches, and its teeth were five inches long. The teeth of *Carcharodontosaurus* contained a series of grooves. Scientists have yet to discover the purpose of the grooves on the teeth.

Why was *Oviraptor* given a name that means "egg thief?"

Oviraptor grew to be about six feet long and had a stumpy, toothless break. The jaws of *Oviraptor*, however, were powerful enough to crush objects as hard as bones. These dinosaurs received their name, meaning "egg thief," because they probably stole the eggs of other dinosaurs and ate them.

What did *Afrovenator* look like?

A largely complete skeleton of a dinosaur named *Afrovenator* ("African hunter") was discovered in Africa in the mid-1990s. This thirty-foot long hunter resembled *Allosaurus,* and had huge jaws studded with two-inch blade-shaped teeth. The *Afrovenator* skeleton also reveals powerful forelimbs, each armed with three curved claws.

What was the size of the largest known *Pachycephalosaurus*?

The largest known *Pachycephalosaurus* was about eight meters, or twenty-six feet long.

Did all extinct dinosaurs lay eggs?

Yes, as far as we know. Unfortunately, only a very few of the dinosaur eggs that have been found contain embryos. One such dinosaur embryo, of species Oviraptor, is on display at the American Museum of Natural History in New York City. In 1993, paleontologists from that museum, along with scientists from Mongolia, discovered a nest of fossilized eggs in the Gobi Desert. In addition to the embryonic *Oviraptor,* scientists discovered in the nest the tiny skulls of two very young *Velociraptors.* Perhaps the *Velociraptors* placed their eggs in the *Oviraptor* nest with the notion that the *Oviraptors* would hatch the eggs. Another possibility set forth by paleontologists at the museum is that the young *Velociraptors* were learning how to hunt and were killed when they attempted to raid the nest. A third possibility that has been advanced is that the *Oviraptor* parents killed the young *Velociraptors* in order to feed them to their hatchlings.

What ancient lizardlike reptile swallowed pebbles and stones to help it sink quickly into the water when diving for fish and perhaps to help it remain submerged when feeding?

Hovasaurus was a lizardlike creature that lived during late Permian times. Its fossil remains were discovered in Madagascar. Not only did it swallow pebbles and stones for ballast, it had another unusual feature: its tail. Its tail was stiff, flat, and nearly twice the length of the rest of its body and, no doubt, was of great use in navigating through water.

What does the name *Pachycephalosaurus* mean?

Pachycephalosaurus literally translates as "thick-headed reptile." These dinosaurs, which lived toward the end of the Cretaceous period, had skulls that were very thick at the top. So thick was the head of *Pachycephalosaurus* that it appeared to have a dome on the top of its head. The thick skull may well have been used for butting other dinosaurs or for pushing heavy loads. Thus, *Pachycephalosaurus* has been divided into two distinct groups: the high-domed ones and the low-domed ones.

Do we know what color *Stegosaurus* was?

No. We have no idea at all what color skin the various dinosaurs had, but scientists have one clue to consider: Most of the largest vertebrate mammals living today, such as the elephant, whale, or rhinoceros, are dark or gray in color.

What is a *nonavian* dinosaur?

Nonavian means "not a bird." All birds are avian dinosaurs. All dinosaurs that are not birds, therefore, are nonavian dinosaurs.

How did the armored dinosaur known as *Nodosaurus*, or "node lizard," get its name?

The body of *Nodosaurus* was covered from its neck to its tail with narrow, rectangular plates of armor. These armored plates were dotted with nodes or small knotty protuberances; hence, the name *Nodosaurus*.

When were the first full-sized dinosaur sculptures commissioned?

The first full-sized sculptures of dinosaurs were commissioned by Richard Owen. In 1854, he commissioned full-sized dinosaur sculptures to be placed upon the grounds of the Crystal Palace in London. These first sculptures are still in existence; if you go to London, you can see them.

Nearly four hundred thousand schoolchildren visit the American Museum of Natural History in New York City every year. When they see an exhibit of dinosaur skeletons, are the children seeing the actual bones of the dinosaurs or are they looking at casts, or copies, of bones?

The American Museum of Natural History in New York City uses, wherever possible, actual dinosaur bones in its exhibits. However, many other dinosaur exhibits in museums throughout the United States use casts. When you go to a dinosaur exhibit, see if you can find out if you are looking at the genuine fossils or if you are looking at replicas.

PLACES TO GO TO SEE DINOSAUR FOSSILS ON EXHIBIT

1. The American Museum of Natural History in New York City
2. Smithsonian Institution in Washington, D.C.
3. Fernbank Museum of Natural History in Atlanta, Georgia.
4. Museum of the Rockies in Rozeman, Montana
5. Denver Museum of Natural History in Denver, Colorado
6. Royall Tyrrell Museum of Paleontology in Drumhellerm Canada
7. The Forum in Vancouver, Canada

Did you know that there is actually a town in the United States called "Dinosaur?"

The National Park Service of the United States Department of the Interior sponsors Dinosaur National Monument in Dinosaur, Utah. If you and your family should ever travel to the Dinosaur Quarry Visitor Center (located about eleven kilometers, or approximately seven miles, north of Jensen, Utah) you will be able to watch paleontologists at work, excavating bones and fossils of animals that lived millions of years ago. The fossils are slowly being uncovered, but they are not removed from the sandstone cliffs. Visitors can view actual dinosaur bones in their natural settings.

The Dinosaur Quarry, the only one of its kind in the United States, is open every day of the year except January 1, Thanksgiving, and December 25. For further information, contact

> The Superintendent
> 4545 Highway 40
> Dinosaur, UT 81610

Where in the United States is Dinosaur State Park located?

If you live in one of our northeastern states, you might want to take in Dinosaur State Park, located near Rocky Hill, Connecticut. While excavating a new state building in 1966, workmen uncovered nearly fifteen hundred dinosaur tracks. These tracks were reburied for preservation. In 1967, a new excavation revealed more than five hundred additional dinosaur tracks belonging to the *Coelophysis* and other dinosaurs.

The trackway is now a registered national landmark, and the park itself is administered by the National Park Service.

To reach Dinosaur State Park (open every day until 4:30 P.M. except January 1, Thanksgiving, and December 25, travel 1.2 miles east on Interstate 91, then take route 23 in Rocky Hill.

        Dinosaur State Park
        West Street
        Rocky Hill, CT 06067

Why are the fossils of very small dinosaurs so rarely found?

The fossils of very small dinosaurs (remember that some dinosaurs were no bigger than hens or large chickens) are very difficult to discover because the bones of the smaller dinosaurs were quite fragile and, therefore, very rarely preserved.

What is the smallest dinosaur of all time?

Taking into consideration that all birds are dinosaurs, the smallest dinosaur is the bee hummingbird (*Mellisuga helenae*) found in Cuba. The weight of a full grown male hummingbird is less than two grams. Compare that to the weight of the extinct *Brachiosaurus*! The *Brachiosaurus* might have weighed as much as eighty-five tons.

What species of dinosaur was discovered in Thailand?

The species *Siamosaurus*. (Siam was the former name for the country known as Thailand, so *Siamosaurus* is easy to remember.) Named in 1986 from the remains of teeth, *Siamosaurus* was a large, two-legged carnivore that might have eaten fish.

How long did certain dinosaurs live?

It is very difficult for paleontologists to figure out just how long certain species of dinosaurs lived. One major clue is growth rings in teeth and bones. Some growth rings lead to the conclusion that dinosaurs lived to be over 100 years old, or perhaps as long as 120 years. If dinosaurs were cold-blooded, however, they might have lived two hundred years or more. Of course, many dinosaurs died fairly young because of disease, bodily injury, or as prey for other dinosaurs.

What was the largest flying bird ever discovered?

The largest flying bird that we know of so far was *Agentivis magnificens*. Looking like a vulture, *Agentivis magnificens* had a wingspan of nearly twenty-five feet and weighed more than two hundred pounds. How would you like to have a two-hundred-pound bird land on your shoulder?

Are there more dinosaur species or are there more mammals alive today?

Since birds are a type of dinosaur, we can say, in one sense, that there are more dinosaur species alive today than there are mammals.

Have all the dinosaurs been discovered?

No, not at all. Scientists have named and classified some three hundred different species of dinosaurs, but it is estimated that ninety-nine percent of all dinosaurs have yet to be discovered. Paleontologists certainly have a lot to look forward to.

Sauropods, which were the largest land animals that ever lived, were once thought to be swamp dwellers because of their great weight. Were they?

Most likely they were not. They were well designed for walking and did not really need the buoyancy of water to support their weight.

Were dinosaurs well-designed for walking?

Dinosaurs were probably much better designed for walking than any other group of reptiles or amphibians. According to scientists at the Smithsonian Museum of Natural History in Washington D.C., "The sauropod body, for example, characterized by a small head atop an extremely long neck, and by an even longer tail, had legs positioned to serve as vertical pillars to support the most weight at the least cost of energy."

In movies about dinosaurs, the dinosaurs are sometimes heard making all kinds of sounds: screeches, honks, roars, etc. How accurate are these sounds? Is it possible for scientists to reconstruct the sounds emitted by extinct dinosaurs?

The sounds made by dinosaurs in such movies as *Jurassic Park* are examples of the filmmaker's art rather than examples of scientific accuracy. However, according to a report by Malcolm W. Browne in the *New York Times* on March 12, 1996, it may be possible, with the help of powerful computers sometimes used to design nuclear weapons, to create a sound that "may be fairly close to the call of the living animal."

Scientists at the Sandia National Laboratory in Albuquerque, New Mexico, are studying the crest of a rare dinosaur named *Parasaurolophus*. The crest of *Parasaurolophus*, a plant-eating member of the hadrosaurs, or "duckbills," contains a labyrinth of long, curved air passages. Scientists are trying to reconstruct what kind of sound the crest might have produced.

Paleontologists are still uncertain about the use of the crest, but some believe that it might have helped *Parasaurolophus* to produce distinctive calls, perhaps to attract mates, warn its kin of the approach of predators, or socialize with others of its own kind.

In 1981, Dr. David B. Weishampel concluded in his master's thesis: ". . . the crest was an excellent resonator that could emit powerful low-pitched sounds, perhaps comparable to the ultra-low notes by which elephants communicate."

Although we do not know what sounds dinosaurs made, the dinosaurs in Steven Spielberg's film version of *Jurassic Park* do make sounds. How did the sound designer create the dinosaur sounds in that movie?

The sound designers for the movie *Jurassic Park,* Gary Rydstrom and his crew, used up to twenty-five different animal sounds to create a range of noises for each of the dinosaurs we see on the screen. The screech used for *Tyrannosaurus rex* came largely from a baby elephant at Marine World. According to Rydstrom: "It [the baby elephant] only roared once. We had to cut and paste and make different variations on it."

For the *Velociraptor,* the sound crew used hoots of mating tortoises, screams from a dolphin, the sound of a horse snorting, and the hiss from an angry goose.

What is the largest known dinosaur egg yet discovered?

The largest known dinosaur egg is *Macroelongatoolithus xixiaensis*. The longate part of its name refers to the egg's elongated shape. This egg was discovered in China's Xixia Basin. The egg measures eighteen inches long.

In giving names to *oospecies,* or egg species, where the egg lacks an embryo and cannot be identified as belonging to any of the known species of dinosaurs, paleontologists take into consideration the size of the egg and other criteria, such as shape, texture, and the pattern of airholes.

MOVIES THAT FEATURE DINOSAURS
(or creatures that are dinosaur-like)

The Lost World (sequel to *Jurassic Park*) to be released in 1997

Jurassic Park (1993)

The Land Before Time (1988)

Planet of the Dinosaurs (1978), starring James Whitworth

When Time Began (1977)

One of Our Dinosaurs is Missing (1976)

At the Earth's Core (1976)

The Land That Time Forgot (1975), based on the novel by Edgar Rice Burroughs

When Dinosaurs Ruled the Earth (1971)

Dinosaurs . . . the Terrible Lizards (1970)

Dinosaurs! (1960)

The Lost World (1960)

The Beast of Hollow Mountain (1956)

King Dinosaur (1955)

The Beast From 20,000 Fathoms (1953)

Fantasia (1940)

King Kong (1933)

The Lost World (1925)

The Ghost of Slumber Mountain (1919)

Gertie the Dinosaur (1912)

The Prehistoric Man (1908)

Is it possible to buy a dinosaur's egg?

It's possible, but it will probably cost quite a bit of money.

Many dinosaur eggs were discovered in China's Green Dragon Mountains. In 1991, one Chinese farmer found some dinosaur eggs, and villagers were soon selling them for only a dollar apiece. But prices rose quickly. Recently, a nest of ten eggs sold for seventy-eight thousand dollars. In Boulder, Colorado, fossil dealers Florence and Charlie Magovern legally bought dinosaur eggs from China and, in 1996, were selling them for one hundred fifty to fourteen hundred dollars, depending on size and condition.

On the average, how many years could a dinosaur live? Did dinosaurs die young or did the average dinosaur live to a ripe old age?

It is difficult to answer that question in general terms, because there are so many different kinds of dinosaurs, and certainly some dinosaurs lived longer than others. On the average, however, dinosaurs lived for some 75 to 125 years, with many dinosaurs dying young because of predators or natural disasters. It is speculated that some dinosaurs might have even attained an age of three hundred years, but that would be most unusual. Still, for an animal to live one hundred years is remarkable. A domestic cat, for example, lives thirteen to seventeen years. A camel may attain an age of forty-five years. Then there is the black Seychelles tortoise, which holds the record for the longest life span of any modern animal: 160 to 170 years!

What does the term *tail club* mean?

A tail club on a dinosaur was a large knot or clump of bone found at the tip of the tail of some armored dinosaurs, such as *Ankylosaurus* and some sauropod dinosaurs.

Most dinosaurs appear to be slow, heavy creatures. Were any dinosaurs fast runners?

We can cite at least one fairly swift dinosaur: *dromaesaurus,* which also possessed a fairly large brain. It takes its name from the Greek word *dramaios,* which means "fast-running."

How fast could the fastest of the extinct dinosaurs run?

It is not easy to know just how fast extinct dinosaurs could have run. The best evidence that scientists have for estimating the speed of dinosaurs is found in fossilized trackways. Scientists measure the length between footprints left by dinosaurs and attempt to estimate speeds in relationship to body height and other characteristics. Medium-sized meat-eating dinosaurs probably attained speeds of not much more than twenty-five miles per hour.

Large sauropods, however, probably walked only two to three miles per hour, which is certainly slower than the average walking speed of a modern human being.

Did any extinct dinosaur weigh over one hundred thousand pounds?

It's possible. *Titanosaurus,* which roamed India and even parts of Argentina, may have been the heaviest dinosaur of all time. *Titano-*

saurus may have weighed between forty-five and eighty-eight tons, or much more than one hundred thousand pounds. Whatever it weighed, *Titanosaurus* was not light on its feet.

Of what use was the bony frill behind the head of *Triceratops*?

That bony frill may have been used to frighten some potential enemies, since it adds to the fearsome nature of *Triceratops*. On the other hand, it seems likely that the frill behind the head increased the strength of the jaw muscles, thus allowing the ceratopsians to eat tougher plant food.

How strong was the bite of *Tyrannosaurus rex*?

According to research performed by Gregory M. Erickson, a doctoral candidate in biology at the University of California at Berkeley, *Tyrannosaurus* could bite with a force equivalent to three thousand pounds. The *T. rex* bite was, in fact, more than three times as strong as the bite of a lion.

No existing animal can bite as hard as *T. rex* could.

What were *pterosaurs?*

Pterosauria was one of the most successful orders of animals: reptiles that could fly. They lived for some 135 million years. Some of the pterosaurs were as large as airplanes, while other species were as small as sparrows. So far, we know of some eighty-five different species of pterosaurs, and more than three thousand specimens have been collected. Pterosaur fossils have been found on every continent except Antartica. The first pterosaur fossil was found in 1784 by Cosimo Collini, who had once worked as a secretary for the great French writer Voltaire (the author of the classic short novel *Candide*).

Still, there are many mysteries surrounding the evolution of pterosaurs. So far, no fossil has been discovered that can be identified as an immediate ancestor of the pterosaurs.

What was the largest pterosaur?

The gigantic *Quetzalcoatlus* (named after the god of the Aztecs) was the largest of the pterosaurs. It was, in fact, the largest animal ever to take to the air. Some specimens suggest that *Quetzalcoatlus* measured some thirty feet across from wingtip to wingtip. *Quetzalcoatlus,* which lived at the very end of the age of the dinosaurs, most likely thrived in parts of what is now the southwestern United States.

Is there any evidence that birds inherited their nesting instincts from dinosaurs?

There is strong evidence that birds inherited their nesting instincts from dinosaurs. Late in 1995, scientists at the American Museum of Natural History announced the discovery of a nine-foot-long ostrichlike creature found nesting on its eggs. In 1993, scientists had discovered this eighty-million-year-old fossil of *Oviraptor* in the Gobi desert of Mongolia. The specimen was seated atop a clutch of fifteen to twenty-two eggs, and its legs were tucked tightly beneath its body in a manner identical to that of chickens and pigeons.

This discovery provided the first direct evidence that dinosaurs took an active role in the care and protection of their offspring.

David Weishampl of Johns Hopkins University School of Medicine said: "Without any imagination at all you can imagine this critter sitting on a nest."

It has been speculated that a giant sandstorm might have been responsible for capturing this specimen of *Oviraptor* and eggs in the position in which they were found.

What dinosaur had the largest brain?

The honor of having the largest brain, or at least the largest brain cavity in the skull relative to its overall body size, belongs to *Stenonychosaurus*. Scientists who have studied the brain cavity of *Stenonychosaurus* say that its brain was slightly larger than the brain of the large Australian bird known as the emu. In spite of its great size, *Stenony-*

chosaurus was about as intelligent as a modern-day opossum.

Are there postage stamps that feature pictures of dinosaurs?

There are quite a few, in fact. The United States has issued postage stamps highlighting dinosaurs. There are even stamp collectors who specialize in collecting nothing but dinosaur stamps. Below is a list of some dinosaur stamps you might be able to acquire (all prices are only approximate and subject to change.) If you are interested in dinosaur stamps, you might ask a local stamp dealer about them.

ANGOLA FDC 8/16/94 New (4) Maxim. cards $9.00
BRAZIL #2316/19 (2 Prs) Mint $4.00
BULGARIA #3540/45 (6) mint $3.15
BULGARIA #3545A Shtlt (1×6)-Mint......................$3.25
BULGARIA New (6v) Mint $5.00
C.A.R. 872/9 (8v) Mint..$7.00
CUBA 2954/9 (6v) Mint..$7.00
CUBA 7265/71 (7v) Mint Scarce...............................$8.00
EQ. GUINEA (7v) Mint..$8.00
FUJEIRA 7/22/68 MK167/76 10v Mint Perf..........$5.75
FUJEIRA Same 10v Mint Imperf $8.15
HUNGARY 2972/7 FDC Max. card 1 Dinosaur......$7.00
HUNGARY 11/16/90 3263/8 (6v) Mint...................$4.70
HUNGARY Same Specimen O/PO (6v).................$35.00
LIBYA 1096a-e 3/21/83 (5v) Dino Mint $8.00
LIBYA Same on 1 FDC...$12.00
MONGOLIA 1871/77 (7v) Mint................................$4.75
MONGOLIA 1878 S/S (lv) Mint...............................$4.75
MONGOLIA 1912/20 (9v) Mint $10.00
MONGOLIA New 5v Mint $5.00
MONGOLIA Same (5v) Specimen only 50 exist..$45.00
MONGOLIA New S/S (lv)...$4.00
MONGOLIA Same (1) S/S Spec. only 50 exist....$45.00
MONGOLIA Same House of Questa Proof Folder with set imperf & S/S perf & imperf!............................$350.00
MONTSERRAT 8/1/92 794 S/S (1) FDC................$6.00
MONTSERRAT 5/8/94 Aquatic Dino FDC (I) $7.00
NICARAGUA 1617/23 on 2 FDC Rare covers $12.00
POLAND 1307/16 (10v) Mint $3.90
ROMANIA 7/30/93 Art Design FDC's (6) $15.00
SAHARA New (7v) & (1 S/S) Mint.........................$7.00
ST. TOME 664/69 & 670 S/S (6+1v) $20.00
ST. TOME Rare max cards 11/30/82 (6) FDC......$48.00

TANZANIA 382/9 (8v) Mint......................................$5.50
TONGA-NIOAF'OU 108/21 (18v) Specimen O/P $37.00
TONGA-NIOAF'OU Shtlt #123 (15v) Spec. O/P .$29.00
TONGA-NIOAF'OU #123 Same Brumide Prf (1) $75.00
THAILAND 1430/33 FDC (1)................................$8.00
(FDC - First Day Cover)

St. Vincent and the Grenadines recently released a postage stamp (Scott number 2052) that shows an *Allosaurus* attacking a *Triceratops*. What mistake did the stamp designer make?

The two species of dinosaurs were not on the planet at the same time. *Allosaurus* thrived in the late Jurassic period and *Triceratops* lived during the late Cretaceous period. Thus, the species *Allosaurus* had been extinct for nearly seventy million years before *Triceratops* appeared on the scene.

How long have birds been flying? Did they fly before the dinosaurs became extinct?

In 1996, paleontologists announced the discovery of an extinct Spanish bird, which was given the name *Eoalulavis hoyasi*. Its name means "early bird with an alula from Las Hoyas." (An alula consists of a tiny tuft of feathers attached to the front of the wing. It allows the bird to fly slowly without losing lift and thus prevents the bird from falling back to the ground.)

The oldest known extinct bird is *Archaeopteryx,* which is about thirty million years older than *Eoalulavis hoyasi.* But *Archaeopteryx* was most likely a very poor flyer.

According to Luis Chiappe of the American Museum of Natural History in New York, *Eoalulavis* ". . . is the first bird in which we find both the modern design in bones and the modern design in feathers."

The finch-sized fossil was discovered squashed between two slabs of rock in a limestone quarry. Its discovery gives rise to speculation that birds were good at flying some 115 million years ago—that they were flying some 50 million years before dinosaurs became extinct.

What is the earliest known bird with a beak, and where were its fossil remains recovered?

The earliest bird with a beak is known to paleontologists as *Confuciusornis sanctus*. In 1994, fossil remains of this earliest bird were discovered in China; hence, the reference to the great Chinese religious leader and philosopher Confucius, combined with *ornis,* meaning "bird".

Scientists speculate that this bird probably

lived some 147 million years ago, or about 70 million years earlier than the previous oldest known toothless bird, *Gobi-pteryx* from Mongolia.

Confuciusornis was about the size of a bantam rooster and had a long reptilian tail, reminding us once again that birds are direct descendants of reptiles. The "Confucius bird" had no teeth, but did have feathers on its body as well as on its wings.

The Confucius bird provides us with the first direct evidence of a bird with body feathers. The only preserved feathers of *Archaeopteryx* (another prehistoric birdlike creature) are on its wings.

The skeleton of what kind of dinosaur was discovered in the Szechuan province of central China in 1978?

The dinosaur discovered in Szechuan in 1978 was named *Yangchuanosaurus*. It was an allosaur with a huge head, a short thick neck, and powerful jaws equipped with sharp fangs. The skeleton is on display today in Beijing's Natural History Museum.

***Tyrannosaurus rex* was an extremely heavy biped. What would have happened to it if it were running along, say, at a speed of forty-five miles per hour and tripped? Could it have survived such a fall?**

Most likely it would not have survived. Paleontologist James Farlow and physicist John Robinson, both of Indiana University, have pondered the scenario of a running *Tyrannosaurus rex* taking a hard fall. According to their calculations (as reported in the April 1996 issue of *Discover*), if *T. rex* were running at top speed (say forty-five miles per hour, although it is not likely that *T. rex* ever ran quite that fast) and tripped, the results would be quite nasty. Since the short forelimbs would not be strong enough to break its fall, *T. rex* would have landed on its stomach with a force of nearly eight thousand pounds. Its head and skull would strike the ground with a force of more than thirty thousand pounds. Such a fall would squash its internal organs. Or, if the dinosaur fell and then skidded on its belly, it would most likely snap its neck when the weight of its body overtook its head.

Since such a fall would have been fatal, *T. rex* probably did not run much faster than fifteen miles per hour. Says Dr. Farlow: "Part of what makes running such a hazardous thing for a beast is simply that if it's going that fast, it has a greater chance of falling. Even if it could [run faster], there are consequences that it might not want to deal with."

How did the species *Dimetrodon* get its name?

The name *Dimetrodon* mens "teeth of two sizes" (the suffix *-don* in animal names frequently refers to tooth or teeth). This dinosaur received its name because its teeth were very sharp and fierce-looking. Some scientists speculate that *Dimetrodon,* which probably fed on small reptiles, may have been the dominant land hunter of its day.

Did dinosaurs see the world in color or in shades of black and white?

According to Gregory S. Paul ("The Art of Charles R. Knight," *Scientific American*, (June 1996), "Dinosaurs . . . probably had color vision like reptiles and birds."

Has the skin of any dinosaur ever been preserved?

Not the skin itself. In 1991, a graduate student named Spencer Lucas was studying rocks near Deming, New Mexico. He discovered a ten-foot long, two-foot wide piece of pockmarked rock that appeared to contain impressions of fossilized tree bark. Five years later, Dr. Lucas, who now works at the New Mexico Museum of Natural History and Science, announced that the rock really contains the fossilized impression of a duck-billed dinosaur's skin. The fossil permits paleontologists to examine and touch an impression of the skin of a dinosaur. There have been about a dozen fossilized duck-bill skin impressions discovered worldwide.

According to Martha Mendoza, writing for the Associated Press, "The impression's texture resembles a mountain bike tire—rough, thick and bumpy, with somewhat symmetrical clumps of little crimped-edged knobs."

What does the Old Testament story of Noah and the Flood have to do with dinosaurs?

If you have read your Bible lately, you might remember the story of Noah and the Flood. At the end of forty days upon the water, Noah sends forth some birds to find out if the waters have abated:

And it came to pass at the end of forty days, that Noah opened the window of the ark which he had made:
And he sent forth a raven, which went

> *forth to and fro, until the waters were dried up from off the earth. (Genesis 8:6, 7.)*

Now what does Noah and his raven have to do with dinosaurs, you ask?

Well, in 1841, when dinosaurs were first identified by scientists as a "Distinct tribe or suborder of saurian reptiles," some people actually believed that the footprints left by dinosaurs belonged to Noah's raven. They thought that Noah's raven walked upon the earth and left enormous footprints. Fortunately, scientists quickly disproved that superstition.

What was the approximate size of the brain of *Stegosaurus*?

Stegosaurus, which had a body about the size of an Asian elephant, had a very small head and an even smaller brain. The size of its brain was about the size of a golf ball or a walnut. The brain weighed all of two ounces.

What was unusual about the legs of *Stegosaurus*?

The back legs of *Stegosaurus* were more than twice as long as its front ones.

Who was Gertie the Dinosaur?

Did you know that one of the first American cartoons featured a dinosaur as its star? She was Gertie the Dinosaur, brought to the silent screen in 1909. (The first film cartoon was *Humorous Phases of Funny Faces,* drawn by James Blackton Stuart and brought to the movie screen in 1906.)

Windsor McKay was a noted cartoonist for the *New York Herald,* the paper for which he drew a comic strip called "Little Nemo in Slumberland." To create the film *Gertie the Dinosaur,* McKay made approximately ten thousand individual drawings. The entire film, however, lasts for only about five minutes.

After the 1909 cartoon *Gertie the Dinosaur* was released, what other early American movies had dinosaurs in them?

Did you know that in 1917, the Edison Film Company (founded by Thomas Edison, the inventor of the light bulb) produced a silent film called *The Dinosaur and the Missing Link*? Four years earlier, however, the great American filmmaker D.W. Griffith had produced another film that featured shots of dinosaurs. *The Primitive Man* (1913) shows Neanderthal men doing battle against large reptilian creatures that look like a cross between a tyrannosaur and a crocodile.

Of course, these early silent films use only primitive animation techniques to recreate the shapes and movements of the large dinosaurs. Although they may have been fun to watch, the films were not very scientific, nor were they accurate.

In what 1925 movie can viewers see the city of London being wrecked by a *Brontosaurus*?

That would be *The Lost World,* a movie adapted from the novel by Sir Arthur Conan Doyle. Another version of that movie, this time produced and directed by Irwin Allen, was made in 1960.

In What 1960 movie does a *Tyrannosaurus rex* fight a steam shovel?

Directed by Irwin S. Yeaworth, *Dinosaurus* is a movie about a caveman, a *Tyrannosaurus rex,* and a *Brontosaurus* brought back to life when they are struck by lightning. In this movie, a *Tyrannosaurus rex* mistakes a steam shovel for another beast and engages it in a fight (a steam shovel does look a bit like a prehistoric animal). A little boy makes friends with the *Brontosaurus* and even gets to ride it. Some of the dinosaur scenes were later reused for the TV series *It's About Time.*

Has there ever been a movie that featured American cowboys fighting dinosaurs?

At least two films have been made that show cowboys shooting at dinosaurs. The first was *The Beast of Hollow Mountain* (1956), starring Guy Madison and Patricia Medina. It was a remake of a 1919 movie, *The Ghost of Hollow Mountain*. The second film, *The Valley of Gwangi,* made in 1969 and directed by James O'Connolly, takes place in Mexico.

In what movie can audiences see a dinosaur take a bite out of the famous Cyclone roller coaster at New York City's Coney Island?

That would be the 1953 Warner Brothers film, *The Beast From 20,000 Fathoms.* That movie is based, at least in part, on Ray Bradbury's classic short story *The Fog Horn.* The dinosaur is one made up by Mr. Bradbury: a rheasaurus.

In the early comic strips, who was Dinny?

Dinny the Dinosaur was perhaps the most famous dinosaur ever to appear in a long-running comic strip. Dinny was a featured character in *Alley Oop,* a comic strip drawn by V.T. Hamlin.

Where is the Lost Continent of Caspak, where explorers can see allosaurs?

The Lost Continent of Caspak is a purely fictional realm created by Edgar Rice Burroughs (the same author who created one of the most famous fictional characters of all time: Tarzan). The Lost Continent of Caspak is featured in Burroughs's novel *The Land that Time Forgot.* One explorer of the Lost Continent sees an allosaur and says, "It looks to me, Whitelu, like an error. . . . Some assistant god who had been creating elephants must have been transferred to the lizard department."

In what novel by Jules Verne is there a description of a fight between an ichthyosaur ("great fish lizard") and a Plesiosaur ("sea crocodile")?

A fight between an ichthyosaur (described in the novel as a hideous monster with "the snout of a porpoise, the head of a lizard, the teeth of a crocodile") and plesiosaur ("a monstrous serpent, concealed under the hard vaulted shell of the turtle") occurs in Jules Verne's classic novel *Journey to the Center of the Earth*. The narrator of that novel describes the fight as follows:

These animals attacked one another with inconceivable fury. Such a combat was never seen before by mortal eyes, and to us who did see it, it appeared more like the phantasmagoric creation of a dream than anything else. They raised mountains of water, which dashed in spray over the raft, already tossed to and fro by the waves. Twenty times we seemed on the point of being upset and hurled headlong into the waves. Hideous hisses appeared to shake the gloomy granite roof of that mighty cavern—hisses which carried terror to our hearts. The awful combatants held each other in a tight embrace. I could not make out one from the other. Still the combat could not last for ever; and woe unto us, whichsoever became the victor.

One hour, two hours, three hours passed away, without any decisive result. The struggle continued with the same deadly tenacity, but without apparent result. The deadly opponents now approached, now drew away from the raft. Once or twice we fancied they were about to leave us altogether, but instead of that, they came nearer and nearer. We crouched on the raft ready to fire at them at a moment's notice, poor as the prospect of hurting or terrifying them was. Still we were determined not to perish without a struggle.

What dinosaur expert, who acted as a consultant for Steven Spielberg's film *Jurassic Park*, published a novel about dinosaurs called *Raptor Red*?

Raptor Red was written by the dinosaur expert Robert T. Bakker. Raptor Red, a young adult female of the species *Utahraptor,* is the main character of this novel written for adults. According to Bakker, "Raptor Red belongs to a species that is making a momentous transition in family life from a male-dominated pack structure to an incipient matriarchy. The adult females have become larger and stronger than the males. They tend to mate for life."

Some of the ideas about *Utahraptor* are only speculation, but the novel provides many stimulating insights into the lives of dinosaurs.

In addition to his novel, Bakker has also written *The Dinosaur Heresies* (Morrow, 1986), a nonfiction work about dinosaurs.

In what novel by Sir Arthur Conan Doyle (the creator of Sherlock Holmes) do the main characters discover a place in South America where dinosaurs still live?

That would be *The Lost World* (a title also used by Michael Crichton in his sequel to *Jurassic Park* based upon the same premise), first published in 1912. In Chapter 10 of Doyle's novel, a newspaper reporter and some professors of paleontology encounter *Pterodactyl:*

> *The place was a rookery of pterodactyls. There were hundreds of them congregated within view. All the bottom area round the water-edge was alive with their young ones, and with hideous mothers brooding upon their leathery, yellowish eggs. From this crawling, flapping mass of reptilian life came the shocking clamor which filled the air and the mephitic, horrible, musty odour which turned us sick. But above, perched upon its own stone, tall, grey, and withered, more like dead and dried specimens than actual living creatures,*

sat the horrible males, absolutely motionless save for the rolling of their red eyes, or an occasional snap of their rat-trap beaks as a dragonfly went past them. Their huge, membranous wings were closed by folding their forearms, so that they sat like gigantic old women, wrapped in hideous web-coloured shawls, and with their ferocious heads protruding above them. Large and small, not less than a thousand of these filthy creatures lay in the hollow before us.

What advanced evolutionary feature did the coelurosaurs (a group of extinct dinosaurs that includes the ornithomimids and the maniraptors) have in common?

Coelurosaurs, including *Teratomis, Deinonychus, Dromiceiomimus,* and *Archaeopteryx,* all had arms that were quite long in relation to their overall body size. Since many other vertebrate animals have also evolved relatively long arms, this is not a unique feature.

Some scientists, though certainly not all, think that coelurosaurs might have used these long arms to capture and tear apart their prey.

The famous creature known as the Loch Ness Monster is supposedly a living dinosaur: a plesiosaur. Does it really exist?

One of the more famous monster photographs ever taken shows the image of the monster as a plesiosaur (a long-necked seafaring reptile). The photograph was supposed to have been snapped by Dr. Robert Wilson, a noted London gyneocologist. In 1994, however, on the sixtieth anniversary of the photograph, two members researching the Loch Ness Monster, David Martin and Alastair Boyd, discovered that the photograph of "Nessie" was nothing but a hoax.

In the 1930's *The Sinclair Book of Dinosaurs*, published by the Sinclair Oil Company, included an anonymous verse about the dinosaur having two brains: one in his head and one in his tail. How does that poem go? Is it true that large dinosaurs had two brains?

The poem is called "The Dinosaur" and it reads:

Behold the mighty dinosaur
  Famous in prehistoric lore,
Not only for his weight and strength
  But for his intellectual length.
You will observe of these remains
  The creature had two sets of brains—
One in his head (the usual place)
  The other at his spinal base.
Thus he could reason "A priori"
  As well as "A posteriori."
No problem bothered him a bit!
  He made both head and tail of it.

So wise he was, so wise and solemn
  Each thought filled just a spinal column.
If one brain found the pressure strong
  It passed a few ideas along;
If something slipped his forward mind
  And if in error he was caught
He had a saving afterthought,
  As he thought twice before he spoke
He had no judgments to revoke;
  For he could think without congestion,
Upon both sides of every question.

Another poet (ahem! the author of this book) once put the matter a bit more succinctly:

> *The dinosaur*
> *Had two brains,*
> *One in his head,*
> *One on his tail,*
> *So how did he fail?*
>
> *Well, let me*
> *Put a bee in your bonnet*
> *If you have a good brain,*
> *Don't sit on it!*

Alas, both verses are based upon an untrue observation.

At one time, it was believed that large extinct dinosaurs had two brains, but that thesis is no longer considered to be true. What passed for a primitive second brain at the tip of a dinosaur's tail was merely a knot of nerves, a ganglia, or an enlarged portion of the spinal cord. *Stegosaurus,* for example, was twenty-five feet in length and had a brain that weighed two and a half ounces. In addition to its tiny brain, the spinal cord in the sacrum was nearly twenty times larger than the brain. This enlarged portion of the spinal cord probably helped to activate nerves in the dermal plates and the long hind legs.

Where can lovers of dinosaurs go to see the tallest dinosaur of all time?

If you travel to Berlin, Germany, and visit the Humboldt Museum (named in honor of Alexander von Humboldt, the great scientist and founder of the modern science of physical geography), you will be able to see a complete skeleton of a *Brachiosaurus grangai* ("arm lizard") that was discovered in Tanzania in 1909.

The dinosaur skeleton is seventy-two feet nine and one half inches in length and is considered to be the tallest dinosaur of them all. It probably weighed somewhere between sixty thousand and eighty thousand pounds.

What is so special about Dinosaur Provincial Park in southern Alberta, Canada?

According to Philip J. Currie, "No area of equivalent size has produced a larger number of articulated dinosaur skeletons of such diversity than Dinosaur Provincial Park in south Alberta." In 1979, the park was designated a UNESCO World Heritage Site. As many as half a dozen new dinosaur skeletons have been found in the park during the past five years.

The species *Albertosaurus,* which lived some 68 million to 76 million years ago, was named in honor of Alberta, Canada.

Why are the La Brea Tar Pits important to paleontologists, and where are they located?

One of the world's best sources of fossils from the Ice Age is located in Hancock Park in Los Angeles, California. The La Brea Tar Pits were formed about a million years ago, when petroleum from oil-bearing rocks seeped to the Earth's surface.

Animals that ventured into the pit were trapped in the heavy, sticky oil and asphalt. The tar perfectly preserved their bones for future generations to uncover.

In 1906, for example, the skeleton of a prehistoric bear was uncovered in one of the layers of tar. Fossils belonging to saber-toothed tigers, giant ground sloths, and many other animals have been discovered in the La Brea Tar Pits; over one million prehistoric remains have, in fact, been recovered so far. Many of the bones and ancient plants recovered from the pits are on display in the George C. Page Museum of La Brea located at 5801 Wilshire Boulevard in Hancock Park, Los Angeles.

Where in the United States is it possible to see replicas of dinosaurs in their natural settings?

If you would like to see replicas of dinosaurs in their natural settings, one place you and your family can travel to is the Dinosaur Gardens and Museum located in Dinosaurland in Vernal, Utah. At Dinosaurland, the gardens are designed to appear exactly as the area did during the age of dinosaurs. In the gardens, fourteen life-sized replicas of dinosaurs are on exhibit. They were sculpted by the well-known artist Elbert Porter. Mr. Porter spent nearly fifteen years working on his sculptures. Some of his replicas are eighty feet long and twenty feet tall! Dinosaurland, in northeast Utah, is located on Highway 40 midway between Denver and Salt Lake City. The gardens are open to visitors all year round.

What does the word dinosaur look like in Chinese?

Have you ever wondered what the word *Dinosaur* might look like in Chinese? Well, you need wonder no longer. Practice doing this and amaze all your friends—except your Chinese friends. They won't be impressed at all.

古生

What prehistoric animal was given its scientific name in honor of a United States president?

Did you know that the scientific name for the Megalonyx (a very large sloth that lived millions of years ago) is *Megalonyx jeffersoni?* Scientists gave it that name to honor Thomas Jefferson, the third president of the United States.

Thomas Jefferson was a very learned gentleman who possessed a great curiosity about a wide range of subjects. In 1799, he gave a talk to the American Philosophical Society. In that talk, he described some large bones and claws that had been found in a cave in Virginia. Jefferson thought that the bones

AMERICAN MUSEUMS OF NATURAL HISTORY WHERE YOU CAN FIND MORE INFORMATION ABOUT DINOSAURS

Academy of Natural Sciences of Philadelphia. 19th and the Parkway, Philadelphia, PA 19143.

The American Museum of Natural History. Central Park West at 79th Street, New York, NY 10024-5194.

Buffalo Museum of Science. 1020 Humbolt Parkway, Buffalo, NY 14211.

Carnegie Museum of Natural History. 4400 Forbes Avenue, Pittsburgh, PA 15213.

Cincinnati Museum of Natural History. 1720 Gilbert Avenue, Cincinnati, OH 45202.

Cleveland Museum of Natural History. Wade Oval, University Circle, Cleveland, OH 44106.

Denver Museum of Natural History. 2001 Colorado Blvd, Denver, CO 80205.

Field Museum of Natural History. Roosevelt Rd. at Lake Shore Drive, Chicago, IL 60605.

Florida Museum of Natural History. Museum Road and Newell Drive, University of Florida, Gainesville, FL 32611.

Michigan State University Museum. West Circle Drive, Michigan State University, East Lansing, MI 48824.

National Museum of Natural History. Smithsonian Institute, 10th Street and Constitution Avenue, NW, Washington, D.C. 20560.

Natural History Museum of Los Angeles County. 900 Exposition Boulevard, Los Angeles, CA 90007.

New Mexico Museum of Natural History. 1801 Mountain Road NW, Albuquerque, NM 87194.

Peabody Museum of Natural History. Yale University, 170 Whitney Avenue, New Haven, CT 06511.

Utah Museum of Natural History. University of Utah, Salt Lake City, UT 84112.

had come from some kind of large lion. Because of the animal's huge claw, he dubbed it *Megalonyx,* a name that actually means "great claw."

Later, scientists discovered that the bones belonged to a hairy slothlike creature that fed on leaves. This prehistoric animal was given the scientific name *Megalonyx jeffersoni* to honor a United States president!

Have any fossils of dinosaurs ever been found in Antarctica?

Yes. In 1986, Paleontologists working for The Argentine Antarctica Institute discovered seventy-million-year-old dinosaur fossils on James Ross Island, northeast of the antarctic peninsula. The expedition was led by Eduoardo Olivero.

The bones were identified as a new plant-eating species within the dinosaur order Ornithischia.

The discovery was of great significance because it verified, for the very first time, the existence of dinosaurs in Antarctica.

How did dinosaurs become extinct?

Once dinosaurs roamed our world. Today they are extinct, which means they no longer exist. What happened? What made it impossible for dinosaurs to go on living?

Throughout the last century, scientists have advanced numerous theories about the extinction of dinosaurs. One of the more recent and widely accepted theories is that a huge asteroid or comet collided with our planet Earth about sixty-five million yeas ago. This collision scattered dust and debris into the atmosphere. The layers of dust cut off great amounts of sun from the earth's surface. For many months it was dark all the time, and the cold was so great that numerous species of plants and animals died. Thus, the dinosaurs, who had reigned for some 160 million years, died out.

Although a comet or asteroid might have struck the earth, many paleontologists believe that a complex of factors—falling sea levels, abrupt changes of climate, and other environmental factors—brought about the dinosaurs' extinction.

If you read newspapers and magazines, you will be certain to find articles pertaining to the demise of the dinosaur.

Where and when was the first reasonably complete skeleton of *Tyrannosaurus rex* found?

In 1902, the bones of a large dinosaur were uncovered in Montana. Three years later, the great paleontologist Henry Fairfield Osborn named it *Tyrannosaurus* ("tyrant reptile"). In 1906, the tyrannosaurids were recognized as a distinct group of theropods.

More people are probably aware of the tyrannosaurids than any other family of dinosaurs. There are four species of tyrannosaurids:

1. *Tyrannosaurus rex*
2. *Daspletosaurus torosus*
3. *Albertosaurus libratus*
4. *Tarbosaurus bataar*

Tarbosaurus bataar was discovered in the Nemegt Basin of Mongolia.

The teeth of *Tyrannosaurus rex* sometimes grew to be as long as seven inches. What happened when *T. rex* lost a tooth? Was the tooth permanently lost, or was *T. rex* able to grow a new tooth in its place?

Tyrannosaurus rex certainly had long teeth, but a member of this species was also lucky in that when it lost a tooth, a new tooth would grow in its place. We also know that there were no dentists in prehistoric times.

What did giant earthworms have to do with the glyptodonts?

Probably not very much. However, an article in a British magazine called *Nature* (February 21, 1878) stated that a German named Fritz Muller, who lived in Brazil, reported on such a giant earthworm. This huge worm was called Minhocao.

The article in *Nature* stated: "About fourteen years ago, in the month of January, Antonio Jose Branco, having been absent with his family eight days from his house . . . on returning home found the road undermined, heaps of earth being thrown up, and large trenches made. These trenches commenced at the source of a brook, and followed its windings, terminating ultimately after a course of from seven hundred to one thousand metres. The breadth of the trenches was said to be about three metres. Since that period the brook has flowed in the trench made by the Minhocao. Paths of the animal lay generally beneath the surface of the stream; several pine trees had been uprooted by its passage."

The anonymous author of the article for *Nature* theorized that the Minhocao was probably not a giant earthworm at all, but that it might have been a very large armadillo, perhaps even "a last descendant of the Glyptodonts"? Glyptodonts were giant armadillo-like creatures that lived during the Pleistocene era.

What dinosaur had teeth that were almond-shaped?

Amygdalodon had almond-shaped teeth. In fact, the Greek root for its name means "almond."

What theory about the extinction of dinosaurs did the American scientist Walter Alvarez set forth in 1979?

After studying the levels of the rare metal iridium found in very old sedimentary rocks in Italy, Walter Alvarez concluded that the earth was struck by a large asteroid or comet some sixty-five million years ago. This comet produced tidal waves and fires, and caused volcanoes to erupt. The comet might also have caused so much dust to be sprayed into the atmosphere that a significant portion of the sun's radiation was cut off from the earth. It was this cataclysm that brought about the extinction of dinosaurs.

What are duck-billed dinosaurs called?

Hadrosaurs is the term scientists apply to those dinosaurs that possessed a ducklike bill.

Can DNA be extracted from dinosaur eggs?

This question was asked in the May 1996 issue of *National Geographic*. The answer was: "Success was reported by a team led by molecular biologist Chen Zhangliang at Peking University. But some scientists remain skeptical. While DNA may well have been isolated, they say, it has not been proved it was from a dinosaur."

Are there any dinosaurs living today?

You may be surprised to learn that there are indeed forms of dinosaurs still living today. There is a group of animals called Archosauria. All dinosaurs are archosaurs. There are two distinct kinds of archosaurs: the Ornithodia and the Crurotarsi. Today's crocodiles and their relatives are Crurotarsi, and all birds belong to the Ornithodia group. Every time you see a crocodile or a bird you are indeed looking at a living dinosaur.

SOME HELP WITH PRONUNCIATION

Below are listed some of the more common dinosaur names that you are likely to come across in your readings. Use the breakdown of the syllables in each name to help you to pronounce the names correctly.

ALLOSAURUS (al-o-*sawr*-us)
BRONTOSAURUS (*Bron*-tuh-sawr-us)
CAMARASAURUS (cam-a-*rah*-sawr-us)
CAMPTOSAURUS (camp-toe-*sawr*-us)
CERATOSAURUS (cer-at-o-*sawr*-us)
CHINDESAURUS (*Chin*-dah-sawr-us)
COELOPHYSIS (seal-o-*fie*-sis)
DIMETRODON (dye-*meat*-row-don)
DIPLODOCUS (dip-*lod*-o-cus)
HADROSAURUS (*had*-ro-sawr-us)
HETERODONTOSAURUS (heh-tuh-ruh-dahn-tuh-*sawr*-us)
IGUANODON (i-*gwan*-o-don)
MEGALOSAURUS (meg-al-o-*sawr*-us)
PTERANODON (tare-*an*-o-don)
SCOLOSAURUS (skol-o-*sawr*-us)

STAURIKOSAURUS (staw-rih-kuh-*sawr*-us)
STEGOSAURUS (*steg*-o-sawr-us)
TRICERATOPS (tri-*ser*-a-tops)
TYRANNOSAURUS (tuh-ran-o-*sawr*-us)

THE GEOLOGICAL TIMESCALE

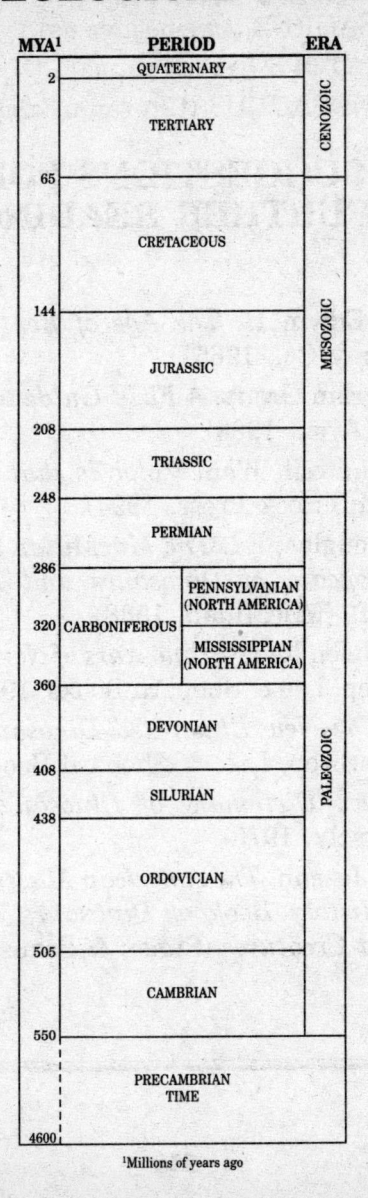

MYA[1]	PERIOD		ERA
2	QUATERNARY		CENOZOIC
	TERTIARY		
65			
	CRETACEOUS		MESOZOIC
144			
	JURASSIC		
208			
	TRIASSIC		
248			
	PERMIAN		PALEOZOIC
286			
	CARBONIFEROUS	PENNSYLVANIAN (NORTH AMERICA)	
320		MISSISSIPPIAN (NORTH AMERICA)	
360			
	DEVONIAN		
408			
	SILURIAN		
438			
	ORDOVICIAN		
505			
	CAMBRIAN		
550			
	PRECAMBRIAN TIME		
4600			

[1]Millions of years ago

SUGGESTIONS FOR FURTHER READING

Colbert, Edwin H. *The Age of Reptiles* (W.W. Norton & Co., 1965).

The Diagram Group. *A Field Guide to the Dinosaurs* (Avon, 1983).

Dingus, Lowell. *What Color is that Dinosaur?* (The Millbrook Press, 1994).

Dixon, Douglas, Ed. *The MacMillan Illustrated Encyclopedia of Dinosaurs and Prehistoric Animals* (MacMillan, 1988).

Sattler, Helen Roney. *Dinosaurs of North America* (Lothrop, Lee & Shepard Books, 1981).

———. *The New Illustrated Dinosaur Dictionary* (Lothrop, Lee, & Shepard Books, 1990).

The Visual Dictionary of Dinosaurs (Dorling Kindersely, 1975).

Wallace, Joseph. *The American Museum of Natural History Book of Dinosaurs and Other Ancient Creatures* (Simon & Schuster, 1994).